Water
77

胃里的塑料

Plastics in My Tummy

Gunter Pauli

[比] 冈特·鲍利 著

[哥伦] 凯瑟琳娜·巴赫 绘

隋淑光 译

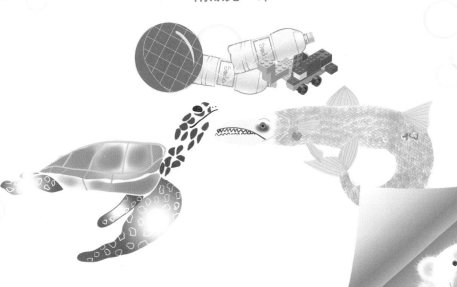

上海远东出版社

丛书编委会

主　任：田成川

副主任：何家振　闫世东　林　玉

委　员：李原原　翟致信　靳增江　史国鹏　梁雅丽
　　　　任泽林　陈　卫　薛　梅　王　岢　郑循如
　　　　彭　勇　王梦雨

特别感谢以下热心人士对童书工作的支持：

匡志强　宋小华　解　东　厉　云　李　婧　庞英元
李　阳　刘　丹　冯家宝　熊彩虹　罗淑怡　旷　婉
杨　荣　刘学振　何圣霖　廖清州　谭燕宁　王　征
李　杰　韦小宏　欧　亮　陈强林　陈　果　寿颖慧
罗　佳　傅　俊　白永喆　戴　虹

目录

胃里的塑料   4

你知道吗?   22

想一想   26

自己动手!   27

学科知识   28

情感智慧   29

艺术   29

思维拓展   30

动手能力   30

故事灵感来自   31

Contents

Plastics in My Tummy   4

Did You Know?   22

Think about It   26

Do It Yourself!   27

Academic Knowledge   28

Emotional Intelligence   29

The Arts   29

Systems:
Making the Connections   30

Capacity to Implement   30

This Fable Is Inspired by   31

一条小梭鱼正在为

几天来的肚子疼而抱怨，他的

爸爸束手无策，只能在海底游来游去，

试图找出造成他儿子难受的原因。这时他注

意到了一只正在转圈的海龟。

梭鱼爸爸问道："打扰一下，你为什么要转

圈呢？"

"哦，幸亏碰到你了，你能帮我弄掉

这根缠住我的渔线吗？"海

龟说道。

A young barracuda
is complaining that his tummy
has been very sore for quite a few
days now. His dad is desperate to
help him and looks around the seabed for
any clues to what could have caused his
son's distress. He spots a turtle swimming in
circles.

"Excuse me, why are you swimming
in circles?" asks the dad.

"Oh, thank goodness you came
along," says the turtle. "Can you
please help me get rid of this
fishing line I'm stuck
in?"

4

你为什么要转圈呢?

Why are you swimming in circles?

让我看看能不能帮你摆脱它

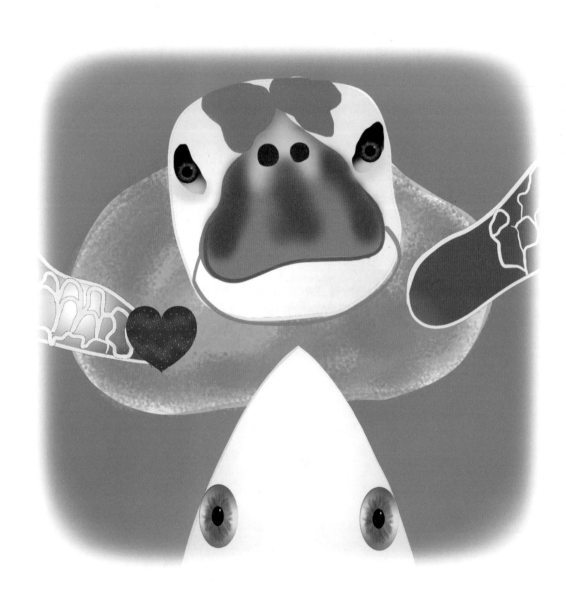

Let me see if I can set you free

梭鱼

爸爸回答道："什
么渔线？我什么都没看到。"

"嗯，人们把渔线做成透明的，这样
鱼就看不到它了。它很结实，一旦你被它
缠住了，就无法摆脱。"

"让我靠近点看看。"梭鱼爸爸回答道，"是
的，我看到了，这条线割伤了你的脖子
和腿。这一定很疼！让我看看能
不能帮你摆脱它。"

"What fishing line? I
can't see anything," replies
the barracuda.

"Well, people make fishing line
transparent so fish cannot see it. They
also make it so strong that once you get
caught in it, there's no escape."

"Let me take a closer look," says the
barracuda. "Yes, I see the line wound
around your neck and your legs. That
must hurt! Let me see if I can set
you free."

"谢谢你。"海
龟说，"这些线感觉比钢丝
还结实，被它缠住后，我既不能游
泳又不能捕食。"
"这真是生死攸关！我尽全力来
帮你。"
"你觉得你能咬断它吗？"
海龟问道。

"Thank you," says
the turtle. "These lines
feel stronger than steel, and
tangled up like this, I can't swim
or feed."
"This is a matter of life and death
then! Let me do the best I can."
"Do you think you will be able
to bite through all this?"
asks the turtle.

被缠住后，我既不能游泳又不能捕食

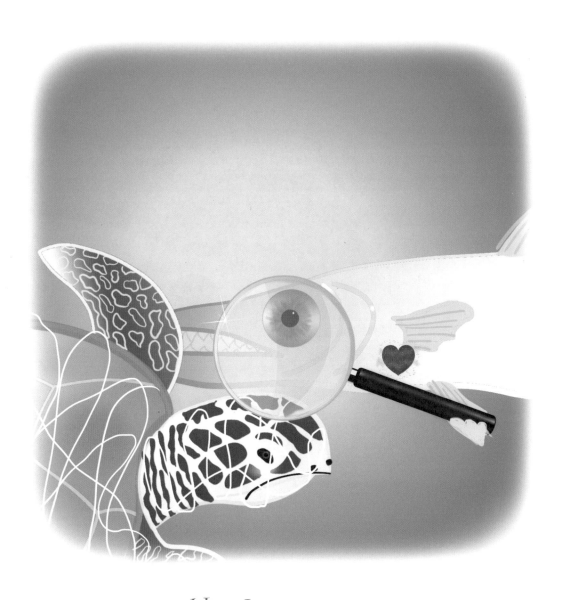

Tangled up like this, I can't swim or feed

幸运的是，我有非常锋利的牙齿

Fortunately, I have very sharp teeth

梭鱼爸爸说："幸运的是，我有非常锋利的牙齿。"然后他迅速咬断了渔线。海龟自由了，浮出海面大口呼吸。

"太感谢了！"海龟说道，"不过请告诉我，渔线的味道怎样？"

"它不像任何其他东西，倒像是人们倾倒进海里的尼龙和塑料。"梭鱼爸爸回答道。

"Fortunately, I have very sharp teeth," says the barracuda, quickly snapping the lines, setting the turtle free to swim to the surface for air.

"I'm so grateful, thank you," says the turtle. "But tell me, how does it taste?"

"It doesn't taste like anything at all, just like the nylon and plastics that people dump in the sea," replies the barracuda.

"我不理解人们

为什么不把海里的渔线、塑

料垃圾、破渔网收集起来，我听说

这些能用来生产服装，扩大就业！"海

龟评论道。

"生产服装？我想人们更喜欢种棉花

来获取服装原料……可种棉花需

要大量的水和农药。"

"I don't understand
why people don't collect
all their fishing lines, plastic
garbage, and broken nets from the
sea. I've been told it can be used to
make clothing and generate jobs!"
remarks the turtle.

"Clothing? I thought people are
sticking to growing cotton for
that … cotton that needs a lot
of water and chemicals
to grow."

生产服装，扩大就业！

make clothing and generate jobs!

你是不是消化不良？

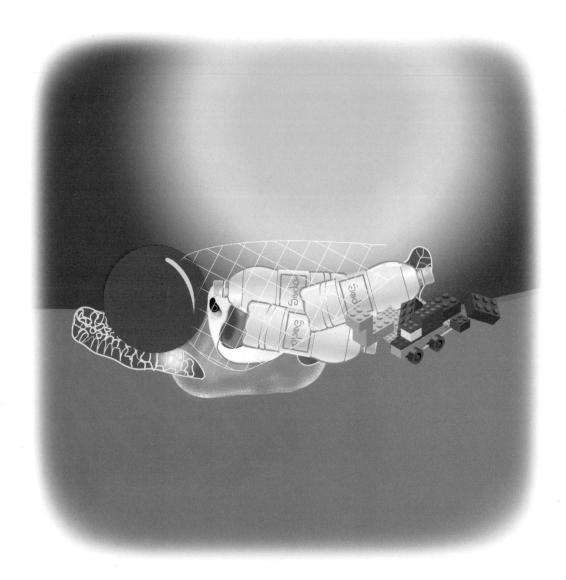

Are you not digesting your food well?

"嗯，一些人已经
意识到要节约用水，并且很快
就可以用其他材料，比如海藻、海草来
生产纺织品了。"
"好吧，陆地上的人面临的问题让他们自己去
应对吧，我们要想办法解决海里的问题，比
如说我儿子经常肚子疼。"
"你看上去也不太好，很臃肿。你
是不是消化不良？" 海龟
说道。

"Well, some people
have realised that they need
to save water and now textiles will
soon be made of other materials, like
algae and seaweed."

"Oh well, let them worry about the
problems they have on land. We have to find
ways to deal with our problems here in the
sea, like my son's frequent tummy aches!"

"You don't look that well either,"
remarks the turtle. "You are very
bloated. Are you not digesting
your food well?"

"我的胃堵得厉害，好像不能正常地进食和排泄了。"梭鱼爸爸叹息道。
"你是不是吃了漂在海里的塑料碎片？"
"塑料？你是说那些在阳光下亮闪闪的，看上去很诱人的东西是塑料？天啊，我要告诉孩子们离它们远点。"

"My stomach feels blocked. It seems nothing that goes in, comes out again," sighs the barracuda.
"Have you been eating those pieces of plastic that float around the ocean?"
"Plastic? You mean those flickering things that look so appetising in the sunlight are bits of plastic? Oh dear, I will tell my children to stay away from it."

你是不是吃了那些塑料碎片?

Have you been eating those pieces of plastic?

人们为什么不用可降解的生物塑料？

Why can't they use degradable bioplastics?

"你知道吗？"

海龟说，"现在是时候完全停
止生产塑料袋和那些会堵塞我们肠道
的高吸水性材料了。人们为什么不用可降
解的生物塑料代替塑料袋来装垃圾呢？"
"我担心的是，在人们变得明智之前，
我的家人和我恐怕还要遭受几年消化
不良的痛苦。"梭鱼爸爸叹
息道。

"Didn't you know?"
asks the turtle. "It's about
time people stopped making
plastics bags altogether. And those
super absorbent materials they block
our intestines too. Why can't they use
degradable bioplastics to collect their
waste, instead of plastic bags?"
"I'm afraid my family and I may still
suffer from indigestion for years
before people become that smart,"
sighs the barracuda.

"嗯，别绝望，朋友。已经有人在用蓟生产塑料了。"

"蓟？我过去以为那是一种杂草！我知道有些真正有创造力的人非常关心环境和我们的生存，他们取得的成绩非常惊人。"

"就像你今天救了我一样惊人！"

……这仅仅是开始！……

"Well, don't despair, my friend. There's already someone making plastics from thistles."

"Thistles? I thought that was a weed! I know there are some truly creative people who deeply care for the environment and our survival, and what they have achieved is truly amazing."

"As amazing as you saving my life today!"

... AND IT HAS ONLY JUST BEGUN!...

……这仅仅是开始！……

… AND IT HAS ONLY JUST BEGUN! ..

For every three tonnes of fish in the oceans, it has been estimated that there is one tonne of plastics. These accumulate into gyres of trash that disintegrate very slowly.

据估计，海里每3吨鱼就对应1吨塑料，这些塑料汇聚形成许多个垃圾漩涡，降解得非常缓慢。

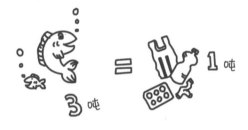

After decades of pollution, some 100 to 150 million tonnes of waste have accumulated as floating plastic islands, often made up of small plastic particles suspended in the sea, called "plastic soup". Every year between 5 million and 12 million tonnes are added.

经过数十年的污染，共有大约1亿—1.5亿吨的垃圾汇集成了塑料浮岛，并且规模以每年500万—1 200万吨的速度在扩大。塑料浮岛通常由小塑料颗粒组成，漂在海面上，人们称其为"塑料汤"。

The size of these huge plastic islands is estimated to have grown to 15 million km². It is expected to double in size in a decade.

据估计，巨型塑料浮岛的面积已经达到1 500万平方千米，预计在未来十年中还会增大一倍。

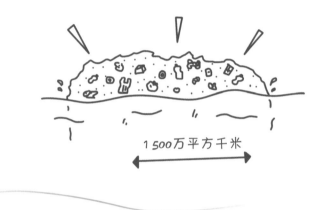

1 500万平方千米

The concentration of small plastic particles in this gyre is up to seven times larger than the concentration of zooplankton.

垃圾漩涡中塑料颗粒的浓度已经超过浮游动物的浓度7倍以上。

Long-lasting plastics end up in the stomachs of marine mammals, fish, and birds. One third of albatross chicks die due to plastics ingested when fed by their parents. The surviving marine life suffers from hormone disruptions due to the additives.

不易降解的塑料最终会进入海洋哺乳动物、鱼和鸟的胃里。有三分之一的信天翁幼鸟因为父母喂食时混入了塑料而死亡。那些幸存的海洋生物则正在遭受由塑料中的添加剂引起的激素紊乱。

When washed down the drain, microbeads from cosmetics and toothpastes pass through sewage treatment plants unfiltered and cause particle water pollution everywhere, from the Great Lakes of North America and elsewhere to the oceans of the world.

牙膏和化妆品里的微小颗粒被冲进下水道，这些无法被污水处理系统过滤的微粒在各地造成"颗粒水污染"，从北美洲五大湖到全世界的海洋都不能幸免。

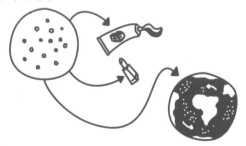

曾经被视作杂草的蓟在闲置的土地上生长得尤其繁茂。作为一种可再生资源，它的花、茎和根能被转化成生产可生物降解化学品的原材料。

Thistles, which grow prolifically, especially in fallow fields, have previously been considered a weed. As renewable resources, the flower, stem, and root can be converted into raw materials for biodegradable chemicals.

荷兰是世界上第一个计划禁止在化妆品中添加塑料微珠的国家，意大利是第一个禁止使用不可生物降解塑料袋的国家，米兰是世界上通过使用生物塑料袋收集垃圾，从而恢复生物量最多的城市。

The Netherlands was the first country in the world to plan doing away with microbeads in cosmetics and Italy was the first country to ban non-biodegradable plastic bags. Milan is the city that recovers most biomass in the world by using bioplastics to collect waste.

Would you like to eat fish that have been feeding on plastics?

你愿意吃以塑料为食的鱼吗?

Do you think that recycling plastics is enough to solve the problem of plastic-waste gyres in the sea?

你觉得通过回收塑料能解决海洋中的塑料废物漩涡问题吗?

蓟这样的杂草能用来生产在阳光、水和土壤中降解的塑料,这是不是很令人惊叹?

Do you find it amazing that weeds, like thistles, are used to make plastics that degrade in the sun, in water, and in the soil?

If you or any of your family members are sick, are you likely to be worrying about other people's problems?

如果你或你的家庭成员生病了,你还有心思去关心其他人吗?

Have a look at everything you have in your home, and that you use on a daily basis, that is made of plastic or contains plastics. Be on the lookout for especially the small plastic particles, known as the microbeads (mainly found in your mother's cosmetics and in toothpaste). Make a list of these items and figure out how these plastics get from your toothpaste or sunscreen into the sea. Now make a drawing that offers a detailed flow chart showing how these plastics pass from your mouth or face all the way into rivers and the sea.

看一下家里和日常生活中由塑料构成或者含有塑料的每一样东西。寻找被称作"微珠"的小塑料颗粒（主要存在于牙膏和你妈妈的化妆品里）。列一张清单，并搞清这些塑料是如何从防晒油和牙膏进入海洋的。现在画一张详细的流程图，标明这些塑料是如何从你的嘴里或脸上出发，最终进入河流和海洋的。

学科知识
Academic Knowledge

生物学	梭鱼会把闪闪发光的物体误当成猎物；塑料降解为小颗粒后被小的水生生物摄入，并在体内聚集，塑料通过这种方式进入了食物链；海龟需要浮出水面呼吸氧气，当它被钓线或渔网困住时就会死去；褐海藻是最好的天然纤维生产者。
化 学	塑料浸泡在水里不到一年就可以降解，释放出双酚A、多氯联苯(PCBs)等有毒化学物质；弹性体是橡胶中的一种添加剂。
物 理	塑料光降解的最终目标是降解到分子水平，但是对大多数塑料来说，仅能分解成小颗粒，仍保持完整的聚合物状态；涡流是一种环形或螺旋形的运动形式；单丝具有较低的光密度，因此不容易被看到；超强吸水聚合物能吸收相当于自身重量500倍、自身体积30—60倍的水。
工程学	渔线由合成纤维制成，即使是由单丝制成的细渔线，也比钢丝坚韧10倍；在热、氧和光的作用下，塑料会随着时间的推移而降解。
经济学	塑料产品的半衰期长达数十年甚至数百年，而其用途只有一天或几周；数十亿计的免费塑料袋仅仅用过一次就被扔掉了；回收高性能的尼龙并将它转化为高性能的服装生产材料是一种共同的社会责任；一次性用品提高了食品的卫生程度但导致了过度包装。
伦理学	塑料只能被风化而不能完全降解，当类似产品的使用寿命结束后，我们怎么可以对其引发的问题放任不管？我们希望由其他人来解决我们自己造成的问题吗？我们生产和使用塑料，在污染地球的同时，最终也会使自己受害，为什么我们不生产耐用的产品来取代一次性用品呢？
历 史	1988年，人类首次发现了海洋中的塑料废物洋流；一次性尿布是在瑞典政府的资助下，于1955年发明的。
地 理	北太平洋环流、南太平洋环流、北大西洋环流、南大西洋环流、印度洋环流；产品包装艺术是从亚洲文化（尤其是日本文化）中发展出来的。
数 学	每平方千米海洋中分布着8平方米的塑料颗粒，其粒径为5毫米×5毫米×1毫米。
生活方式	现代社会已经成了一次性用品社会；尿布的舒适度取决于超吸收性能的化学材料对尿的吸收能力，这种材料不能降解，只会在水中被侵蚀。
社会学	社会上许多人把一次性用品视作提高生活质量的手段；包装产品和打开包装的乐趣。
心理学	由于认知上的错误，人们既没有回收塑料垃圾，也没有想办法避免或补救这一问题，需要引入经济激励机制，或者采取征税这样的抑制措施来提高塑料回收效率。
系统论	未经充分考虑的、不成功的产品设计会降低生活质量，并从根本上影响生活。

情感智慧
Emotional Intelligence

梭 鱼

梭鱼一家正在遭受病痛，但他们并不知道原因。尽管这样，梭鱼爸爸还是很关心别人，当他发现海龟行为异常时表现出了同情。但是他并不同情人类。梭鱼爸爸不知道渔线是什么，但是一旦意识到海龟面临生命危险，他立即准备用牙齿解救她。梭鱼爸爸不相信人类会欢迎那些真正有利的发明，因此他仍然想继续关注海洋所面临的问题，而不是去考虑陆地上人们所面临的抉择。当海龟提到塑料的替代品，如用蓟生产的生物塑料时，梭鱼爸爸发现这超出了他的理解范围。

海 龟

海龟被渔线缠住了，处于生死存亡的紧急关头，梭鱼爸爸伸出援手解救了她，对此她非常感激。在指挥梭鱼爸爸救她时，海龟表现得既平静又有自控力。被解救出来后，她表达了谢意，然后询问塑料的味道，并且对人们不采用已知的方法来解决塑料污染问题感到不理解。海龟非常清楚人类所面临的挑战和抉择。她关注梭鱼一家的健康情况，并指出问题是因为误食塑料引起的。尽管梭鱼对她提到的新发明不抱信心，但海龟出于对救命之恩的感激，还是提供了更深入的信息，分享了自己的智慧。

艺术
The Arts

塑料垃圾通常被认为是废品，但可以回收再利用。试试能不能收集25片废塑料，并用来制作某种形式的艺术品。比如说用从塑料提袋上拆下的塑料条编成小垫子；从塑料瓶上切下圆片，来做一件时尚的衣服。邀请你的家人和朋友一起收集废塑料，然后探讨可以把它们做成什么形式的艺术品。只要实用就可以了，不一定非得华丽。

思维拓展
Systems: Making the Connections

　　给我们带来便利的塑料产品已经成为现代生活的一部分。在食品包装方面，它们由于更便宜、更卫生，正在逐步替代金属产品。一次性塑料购物袋也已经取代了纺织品购物袋这样的耐用品。但是，当塑料产品的优势越来越明显时，我们对它的危害却视而不见。如此多的不可降解的塑料产品（其半衰期即使没有几百年也有几十年）遍布于我们的生活中，正在渐渐汇集成一个沉重的负担。人们努力推动塑料回收，并纠正填埋和焚烧的过激做法。然而我们的环境中仍有大量塑料，遍布街道、山坡、峡谷、河流、海洋。尽管我们早就被警告过塑料不易降解，但只有在得到令人信服的统计数字后，在看到海龟、鱼和鸟死亡的令人不安的照片后，这个问题的严重性才凸显出来。有毒的化学品正在进入食物链，威胁着一切生命。塑料曾经被视为现代化的标志，现在却成为一个令人不安的现实问题。人们必须重新设计塑料，使其既保留功能，又不会对土地、水生生物以及我们赖以生存的食物链造成威胁。目前新的产品已经出现了，它的意义已经超越了用一个产品代替另一个产品，而是提示人们要选择更合适的经济发展模式。将蓟作为一种可再生资源来生产生物塑料或者其他复合物，这不仅宣告了无污染产品设计的出现，而且还把容易获得的资源转变为经济增长和扩大就业的机遇。

动手能力
Capacity to Implement

　　你在家里和学校使用过多少种生物塑料产品？看一下当地商店使用的塑料袋和食品包装物是不是用可再生材料做成的。然后研究一下这些材料在阳光下、水中和土壤中进行生物降解的可行性数据。列出可以用生物塑料取代的塑料清单。也去车库看看汽车所用的润滑油，确认一下是否只是用石油生产的。一旦你列出了所有可以用生物塑料替代的产品清单，看看哪个供应商保证他们提供的塑料产品和润滑油是用可生物降解的可再生资源生产的。现在你可以制订一个促进可持续发展的日程。

故事灵感来自
This Fable Is Inspired by

卡狄亚·巴斯蒂奥利
Catia Bastioli

　　卡狄亚·巴斯蒂奥利是一位科学家出身的意大利企业家。在佩鲁贾大学获得化学学位后，她作为一个材料学家，在蒙特爱迪生化工集团开始了研究生涯，致力于从可再生资源中寻找可生物降解的塑料。当公司所遵循的传统商业模式失败，业务解体后，她把她的研究部门转变为一个独立单位，然后转型为一个叫诺瓦蒙特的新企业。她继续致力于用一系列可生物降解和合成的生物塑料（被称为Mater-Bi）来发展生物化学工业，这是一种不会与食物生产相竞争的生产方式。卡狄亚·巴斯蒂奥利及其团队在生物塑料研制以及基于本地的原材料（包括农业废弃物和杂草）规划当地经济发展方面取得了进展。她的努力促进了产业的进一步转型，把无效的生产单位转变成有效的生产单位。卡狄亚·巴斯蒂奥利拥有超过100项与生物聚合物有关的专利，证明人们通过重新设计塑料购物袋，可以实现创新，从而为海洋和陆地的问题提供新的解决思路。2007年，卡狄亚·巴斯蒂奥利荣获"欧洲发明者"称号。

图书在版编目（CIP）数据

冈特生态童书.第三辑修订版:全36册:汉英对照 /
(比)冈特·鲍利著;(哥伦)凯瑟琳娜·巴赫绘;
何家振等译.—上海:上海远东出版社,2022
书名原文:Gunter's Fables
ISBN 978-7-5476-1850-9

Ⅰ.①冈… Ⅱ.①冈…②凯…③何… Ⅲ.①生态环
境–环境保护–儿童读物—汉、英 Ⅳ.①X171.1-49

中国版本图书馆CIP数据核字(2022)第163904号
著作权合同登记号图字09-2022-0637号

策　　划　张　蓉
责任编辑　祁东城
封面设计　魏　来李　廉

冈特生态童书
胃里的塑料
[比]冈特·鲍利　著
[哥伦]凯瑟琳娜·巴赫　绘
隋淑光　译

记得要和身边的小朋友分享环保知识哦！
八喜冰淇淋祝你成为环保小使者！